Housing
120

# 捉迷藏

## Let's Play Hide and Seek

### Gunter Pauli

[比] 冈特·鲍利　著

[哥伦] 凯瑟琳娜·巴赫　绘

唐继荣　张晓蕾　译

上海远东出版社

# 丛书编委会

主　任：田成川

副主任：闫世东　林　玉

委　员：李原原　祝真旭　曾红鹰　靳增江　史国鹏

　　　　梁雅丽　孟小红　郑循如　陈　卫　任泽林

　　　　薛　梅　朱智翔　柳志清　冯　缨　齐晓江

　　　　朱习文　毕春萍　彭　勇

特别感谢以下热心人士对童书工作的支持：

匡志强　宋小华　解　东　厉　云　李　婧　庞英元

李　阳　梁婧婧　刘　丹　冯家宝　熊彩虹　罗淑怡

旷　婉　王靖雯　廖清州　王怡然　王　征　邵　杰

陈强林　陈　果　罗　佳　闫　艳　谢　露　张修博

陈梦竹　刘　灿　李　丹　郭　雯　戴　虹

# 目录

# Contents

一条饥饿的金枪鱼在水里扑腾着，四处寻找食物。它四处张望，但没有找到一点吃的。一只海蛞蝓注意到了这条绝望的金枪鱼，问道："你在担心日本渔民吗？"

　　"日本渔民？我为什么要担心他们？我只是在担心找不到食物。"

A hungry tuna fish is swimming up and down, looking around for food. It looks everywhere but cannot find a thing to eat. A sea slug notices the desperate tuna and asks: "Are you worried about the Japanese fishermen?"

"Japanese fishermen? Why would I worry about them? I worry about finding food."

一条饥饿的金枪鱼在水里扑腾着……

A hungry tuna fish is swimming up and down ...

······在玩捉迷藏。

... playing hide and seek.

"海洋里到处都是食物，只是你不一定看得见它们。"

"我知道，周围有食物。不过很明显，它在和我玩捉迷藏。"

"捉迷藏？你开玩笑吧！这是一片汪洋大海，从上万米深的海底到太阳升起的海平面之间，没有任何中间地带，根本什么都藏不住。"

"The sea is full of food. But you are not likely to see any."
"I know, there is food around, but it is clear that it is playing hide and seek."
"Hide and seek! You've got to be kidding. Here in the open ocean, with the sea floor ten thousand metres below and the sun shining on the surface – and nothing in between – there is simply nothing to hide behind."

"我知道，但是我的美餐仍然成功地消失不见了。"

"啊，就好像人类造的有些飞机雷达探测不到一样。"

"对！那些飞机的确是在天上飞，没有任何地方可以躲藏，可我们仍然看不见它们。"

"I know, but still my meal has succeeded in turning invisible."

"Ah, like those planes people make that cannot be seen on a radar screen."

"Exactly, those planes are in the air, and there is nowhere for them to hide – and still they are invisible."

······消失不见了。

... turning invisible.

······在撕咬一条鲨鱼······

... trying to bite a shark ...

"听起来很有趣哇! 它们怎么做到的? " 海蛞蝓问道。

"我也不清楚，但我知道遇到这种情况要非常小心。有一回我满心以为得到了一顿丰盛的午餐——一条很大的鱼，却猛然发现我居然在撕咬一条鲨鱼，而它反过来咬了我一口。"

"That sounds like fun! How do they do it?" Sea Slug asks.
"I have no idea, but I do know I have to be very careful. Once I thought I was going for a hearty lunch, a good size fish, only to realise that I was trying to bite a shark, who instead got a chunk out of me."

"一条隐形的鲨鱼？"

"受了那次惊吓后，我一连好几个星期都做噩梦。现在我变得谨慎多了，但我真想知道我的下一顿饭在哪里。我好饿！"

"好吧，我来告诉你一个秘密。一看到你的影子，你的猎物们就打开了它们的灯。而当你四处张望着游过，却以为那光亮不过是太阳光罢了。"

"An invisible shark?"

"I had nightmares for weeks after that accident. Now, I am more careful. But I do want to figure out where my next meal is. I am hungry."

"Well, let me tell you a secret. A couple of your dinners turn on their lights the moment they spot your silhouette. The moment you swim by and look up, you think that there is nothing but the shining sun."

......一看到就打开了它们的灯......

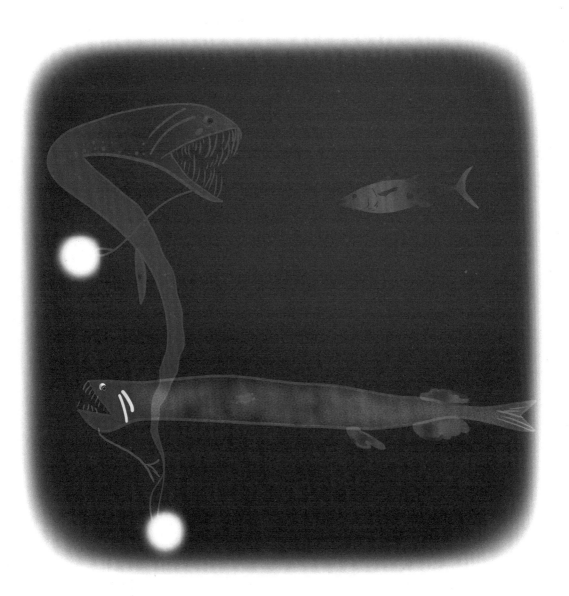

... turn on their lights when they spot ...

......透镜、镜子和滤光片......

... lenses, mirrors and filters ...

"逆光么？瞧，我的食物身上可没有任何电池或电线，也没有连接到电网。"

　　"没错。"海蛞蝓飞快地反驳道，"但它们有透镜、镜子和滤光片，这些东西使它们看起来就像与天空融为一体的蓝色海水。"

"Counter light? Look, my food does not have any batteries or wires, and no connection to the grid."
"Right," the slug quickly counters, "but they do have lenses, mirrors and filters to make themselves look like the blue water blending into the sky."

"所以当我抬头看的时候，它们就隐形了。但是，为什么当我向下看的时候，也看不见它们呢？"

"你真的想知道？"

"拜托，这关乎我的食物和生存问题。"

"So some become invisible when I look up, but what makes them invisible when I look down?"

"You really want to know?"

"Come on, it is about food and survival for me."

当我向下看的时候，也看不见？

... invisible when I look down?

……通过皮肤来吸收营养物质。

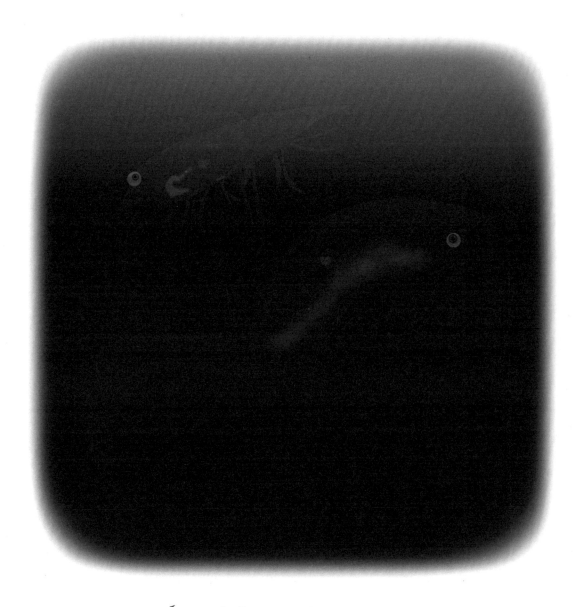

... absorb nutrients through their skin.

"好吧，它们的身体是透明的！"

"这不可能！它们胃里的东西总能被看见吧！"

"它们根本就没有胃。这些家伙是通过皮肤来吸收营养物质的。"

"Well, they have a see-through body."

"That is not possible! Surely what is in their stomachs can be seen!"

"They do not have a stomach. These guys absorb some of their nutrients through their skin."

"既透明又能吸收食物的皮肤？如果它们跟太阳靠得太近，肯定会被晒伤。"

"别担心！它们透明的皮肤甚至会分泌防晒霜。"

"海洋世界的美妙总是让我惊讶！我只是希望今天能饱餐一顿，明天不要成为别人的口粮。"

……这仅仅是开始！……

"A see-through and an eat-through skin? If these get too close to the sun, they will surely get sunburn."

"Don't you worry. They even have suntan lotion in their transparent skin!"

"This wonderful world of the sea never fails to surprise me! I just hope I can get my meal for the day, and avoid being eaten tomorrow."

... AND IT HAS ONLY JUST BEGUN! ...

......这仅仅是开始！......

... AND IT HAS ONLY JUST BEGUN! ...

# Did You Know?

你知道吗?

On land animals camouflage themselves amongst foliage and terrain. In coastal waters, sea creatures blend into coral, seaweeds or rock. But, in the deep ocean creatures have nowhere to hide. Being translucent is therefore the perfect camouflage.

在陆地上,动物利用树叶和地形来伪装自己。在沿海水域,海洋生物藏身于珊瑚、海藻和岩石中。但是在深海,生物无处可藏,因此让自己身体变得半透明是完美的伪装。

The lenses of our eyes are transparent tissue, as light passes through them. Biological lenses are made from protein and perform better than the ones made from glass or polymers.

我们眼睛的晶状体是透明组织,光线能够穿透。生物晶体是由蛋白质构成的,比用玻璃或聚合物制作的晶体性能更好。

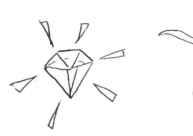

半透明的动物并不会吸收光线或反射自身的颜色。相反，它们的身体能够让光线透过，我们看到的颜色是光线从半透明生物后面的物体反射回来的。

Translucent animals do not absorb light or reflect colour back. Instead, their bodies allow light to pass through and we see colour bouncing off the objects behind the translucent animal.

为了使身躯透明，动物的身体需要吸收或者散射尽可能少的光线。要演示什么是透明度，空气和水是最佳且最充足的材料，而钻石是最珍贵的。玻璃是第一种人造的透明材料。

In order to be transparent the body must absorb or scatter as little light as possible. Air and water are the best and most abundant examples of transparency, while diamonds are the most precious. Glass was the first man-made transparent material.

When light falls onto a material, part of the light is reflected and part of it bends (refraction). That is why there are no transparent cows, as the density of the air is much less than that of flesh.

当光线照射到物体上，一部分光线被反射，还有一部分光线产生弯曲（折射）。因为空气的密度要比血肉之躯小很多，所以世界上没有透明的牛。

The sea slug is a mollusc without a shell. There are thousands of varieties living in coastal zones, as well as in water up to 2,500 metres deep. These animals include some of the most colourful species on earth and use stored toxins from their prey as their defence against predators.

海蛞蝓是一种没有壳的软体动物。海岸带水域中有成千上万种海蛞蝓类生物，分布在水深至 2 500 米下的区域。这些动物中有地球上色彩最艳丽的生物，它们会储存来自猎物的毒素，将其用来抵御天敌。

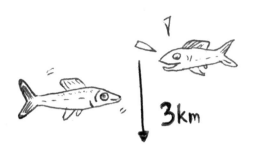

The cookie-cutter or the cigar shark employs counter-illumination with tiny light producing organs providing perfect camouflage, making it look like a small fish. It migrates vertically for 3 kilometres every day.

巴西达摩鲨又称雪茄鲛，它采取一种"反照明"的方式，用微小的发光器官提供完美的伪装，使得它看起来像是很小的鱼。巴西达摩鲨每天的垂直游动距离长达 3 千米。

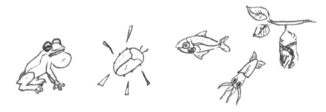

Transparency is not limited to the sea. The glass frog, the golden tortoise beetle, the transparent Amazonian fish, larval squid and the monarch butterfly pupa are all translucent too.

生物透明现象不只在海中才有。玻璃蛙、黄金龟甲虫、透明的亚马孙鱼类、鱿鱼幼体和黑脉金斑蝶蛹都是半透明生物。

# Think About It

## 想一想

Would you enjoy the power to become invisible to friends and family?

在家人和朋友面前隐身，你享受这样的能力吗？

Sharks and whales can swim around in the ocean without fear, but how would you feel not being able to hide from anyone?

鲨鱼和鲸可以无所畏惧地在大海中畅游，但如果你在海里面对危险无法躲避的时候，你会有何感想？

How would you enjoy Nature, if birds and the bees were invisible?

如果鸟类和蜜蜂都隐身了，你该如何欣赏大自然的美景呢？

Think about swimming in the sea, and having the ability to turn on dozens of little lights that would shine towards the bottom, so that anyone looking up would not be able to see you. Would you feel strong? Would you feel smart?

试想一下，如果你在海里游泳时可以打开几十盏小灯照向底部，使得任何往上看的生物都不会发现你。你会觉得自己很强大、很机智吗？

# Do It yourself!
# 自己动手!

One of the ways to become invisible is to use mirrors. The silvery side of herring and sardines is a system of mirrors. Go outside on a sunny day with a dozen mirrors on hand. Now create two teams, one holding the mirrors, and the other looking at the mirrors. Get at least 10 mirrors to reflect the sunlight, so that those looking at you are blinded and cannot see you. You have made yourself invisible!

使用镜子是让自己隐身的一种方法。鲱鱼和沙丁鱼的银色侧身就是一套镜子盔甲。在一个阳光明媚的日子，带上十几面镜子到室外去。现在大家分为两组，一组拿着镜子，另一组看着镜子。用至少十面镜子反射阳光，让对面看你的人睁不开眼，因而看不见你。你就成功地将自己隐身啦!

# 学科知识

## Academic Knowledge

| | |
|---|---|
| 生物学 | 一条5岁的雌性金枪鱼产卵500万枚，15岁时则高达4 500万枚；金枪鱼以无脊椎动物为食；金枪鱼其实是有骨头的，并非软骨鱼类；像其他鲨鱼一样，巴西达摩鲨的牙齿也会定期更新；巴西达摩鲨属于伏击型捕食者。 |
| 化 学 | 金枪鱼含盐量低，是很好的维生素和矿物质来源；生物体可以从食谱中获取荧光素，但它们也进化出了自己的酶和发光细胞来控制发光；自然光由生物发光、化学发光、荧光和磷光所致。 |
| 物 理 | 生物发光是生物在深海中的主要交流机制；透明意味着你的视线能完全透过它，而半透明意味着光线能够穿过去；光折射率；反照明；发光器官；折射定律；凹透镜和凸透镜。 |
| 工程学 | 可以放置于眼中的人工晶体；新的眼科手术工程。 |
| 经济学 | 世界上捕捞的蓝鳍金枪鱼80%都在日本出售，一条鱼就价值上百万美元；对濒危物种而言，价格更高的负面影响是，鱼越少价格越高，价格越刺激捕得越多；巴西达摩鲨没有商业价值，因而并不是濒危物种。 |
| 伦理学 | 当96%的金枪鱼渔业资源被消耗掉时，人们为何还在继续捕捞金枪鱼，仍然没有拟定停止捕捞的协议？ |
| 历 史 | "不给糖果就捣蛋"和"化装"的传统起源于中世纪，孩子们在万圣节期间会乔装打扮，挨家挨户歌唱或祷告，来讨要食物或钱财；2008年诺贝尔化学奖得主下村修、马丁·沙尔菲和钱永健从维多利亚多管发光水母身上提纯了钙激活光蛋白。 |
| 地 理 | 海洋覆盖了约71%的地球表面，也代表了地球上超过90%的宜居区域。 |
| 数 学 | 价格弹性的计算：价格的增长怎样影响市场需求的增减，这种增减又反过来影响价格和需求的变化；这种演化怎样受替代品的冲击，该产品被视作必需品还是奢侈品，以及这又怎样影响总收益。 |
| 生活方式 | 按富人的消费水平来消费的欲望导致物种灭绝。 |
| 社会学 | 金枪鱼的英文单词"tuna"来自希腊语，意思是"匆忙奔跑"；一项调查发现，在英国超过一半的房屋主人会在万圣节关掉家里的灯，装作不在家的样子。 |
| 心理学 | "我也一样"效应（"me-too" effect）指我们复制意见领袖的做法、饮食和购买行为；我们知道金枪鱼正走向灭绝，但我们仍然在不断消费它，就好像我们从来不知道它濒临灭绝，这从来影响不到我们，又或是假装自己只是破例一次。 |
| 系统论 | 过去常用来捕捞整群金枪鱼的大网已经导致现在金枪鱼接近商业性灭绝；这些大网也用来捕捞海龟和海豚；金枪鱼位于海洋食物链的顶端，它的资源崩溃会对整个海洋生物系统产生效应。 |

## 情感智慧
### Emotional Intelligence

海蛞蝓

海蛞蝓关心金枪鱼，指出日本人对金枪鱼的胃口不小。海蛞蝓揭示了明摆着的事实：大海中到处都是食物，根本藏不住。海蛞蝓不了解可以隐身的生物，但它把这与自己原有的知识联系起来。海蛞蝓坦言自己也不知道其他海洋生物是如何让自己变得不可见的，但它渴望了解其中缘由。它迅速补充了有些生物是怎样成功隐形的，还细谈了隐身所采用的方法。从顶部看和从底部看视角不同，解释也有差异，但海蛞蝓使出了浑身解数去解释，包括重新设计消化系统。

金枪鱼

金枪鱼对未来并不担忧，它只担心当下。为什么食物在无处可藏的情况下还是看不见？它意识到无论在空中还是深海食物都无处隐身。于是，它为自己神出鬼没的食物感到沮丧，同样，天敌也难以预料。金枪鱼有信心地分享它的亲身经历，由于"灯光技术"而突然出现的鲨鱼是怎样撕咬它的。这一遭遇给金枪鱼造成了创伤。由于仍然局限在原有知识框架，海蛞蝓的解释在金枪鱼看来是不现实的。当金枪鱼理清了一点儿思路后，又迅速质疑一切。它认识到大海仍然充满惊奇，有想把一切弄明白的冲动。

## 艺术
### The Arts

"不给糖果就捣乱"和南瓜雕刻都是古老的游戏。用刀片雕南瓜太过危险，你可以做一个模具在南瓜上饰以花纹。这比雕刻南瓜来得更快，也没那么危险。明智的做法是，先选择合适的南瓜，这也是一门艺术。南瓜的形状应该能保持直立。大南瓜可能容易往一边倒，而中等大小的南瓜刚好合适。现在，找一个表面光滑、没有凹痕和刮痕的南瓜。即便你使用了模具，你仍然可以选择去雕刻，只是要遵循模具的线条来进行。准备好了吗？

## 思维拓展
### Systems: Making the Connections

人类可以定居并且有效利用的文化和生存空间只占地球表面宜居区域的10%。海洋占据了地球表面的三分之二，但生命体遍及所有海水深度。太阳射线和光线无法穿透超过25米深的海水，同时几百米深的海水氧含量很低，深海生物不得不改变自己和适应环境。生存条件由诸如食物、氧气这类基本的需要所决定，而所有生物只能在环境和物种的能力范围内演化。这就导致了猎物和捕食者之间极具挑战性的关系和交流系统，以及看上去就像是来自科幻小说的最巧妙的解决方式。虽然在雷达显示屏上看不到的隐形轰炸机被认为是一场技术革命，但大多数水下物种早已经发展出大量类似的功能。在黑暗中躲藏的能力、对镜子和透镜的利用、消化系统的透明和通过皮肤进行食物吸收的适应，这些都包含大量的灵感，当然也包含创造性的适应和进化经验课程——为了获得期望的效果（食物及生存和生育能力），这些生物只采用在当地能获取的资源，有效利用物理法则以及调整化学世界。然而，对这个未经探索的世界，人类却缺乏敬畏之心。人类的欲望会导致破坏性的行为。即便金枪鱼已至灭绝边缘，但由于其昂贵的价值，它仍然会被猎捕，并用私人飞机运往日本做成寿司和生鱼片。金枪鱼位居海洋食物链顶端，它的消失将对整个生态系统产生巨大冲击。为了创建一个拥有共同栖居文化的世界，赞美大洋的慷慨固然重要，同时也必须克制我们的欲望。不管在陆地上还是海洋中，我们都并行在一条进化路径上，渴望确保我们的生存。这意味着要抑制我们的欲望，尊重与我们共享这个星球的所有其他生物的生命。

## 动手能力
### Capacity to Implement

关注世界上所有的金枪鱼。蓝鳍金枪鱼极度濒危，黄鳍金枪鱼和长鳍金枪鱼也接近濒危状态。拜访附近的餐馆，问服务员要一份菜单。如果看到金枪鱼出现在菜单上，那么看看该餐馆是否供应其他不濒危的鱼类。列出一份包含所有资源供应充足并能提供美味大餐的鱼类清单。去一些超市看看，是否有超市出售金枪鱼。说服它们替代的鱼类也很美味，然后补充说明，只要市场仍然有需求，我们就无法改变过度捕捞的现状。

## 故事灵感来自
## This Fable Is Inspired by

# 桑可 · 约翰松
# Sönke Johnson

桑可 · 约翰松在美国宾夕法尼亚州的匹兹堡长大，而匹兹堡是一座钢铁厂无处不在且夜空被橙色光污染的工业化城市。他先从宾夕法尼亚州的斯沃斯莫尔学院取得数学学位，然后从北卡罗来纳大学查珀尔希尔分校取得生物学博士学位。在过去的 25 年，约翰松研究开放大洋的视觉和伪装现象，也致力于沿海和陆生物种、磁信号接收、夜间照明和人类白内障的研究。同时，他还探索动物所采取的光学和视觉技术的进化和多样性。他在位于北卡罗来纳州达勒姆县的杜克大学担任生物学教授，他在该校创建了自己研究伪装、信令和非人类视觉模式的实验室。

**图书在版编目（CIP）数据**

冈特生态童书.第四辑:修订版:全36册:汉英对照 /
(比)冈特·鲍利著;(哥伦)凯瑟琳娜·巴赫绘;
何家振等译.—上海:上海远东出版社,2023
书名原文:Gunter's Fables
ISBN 978-7-5476-1931-5

Ⅰ.①冈… Ⅱ.①冈… ②凯… ③何… Ⅲ.①生态环
境–环境保护–儿童读物—汉、英 Ⅳ.①X171.1-49

中国国家版本馆CIP数据核字(2023)第120983号
著作权合同登记号图字09-2023-0612号


策　　划　张　蓉
责任编辑　张君钦
封面设计　魏　来 李　廉

冈特生态童书
### 捉迷藏
[比]冈特·鲍利　著
[哥伦]凯瑟琳娜·巴赫　绘
唐继荣　张晓蕾　译

记得要和身边的小朋友分享环保知识哦！
八喜冰淇淋祝你成为环保小使者！